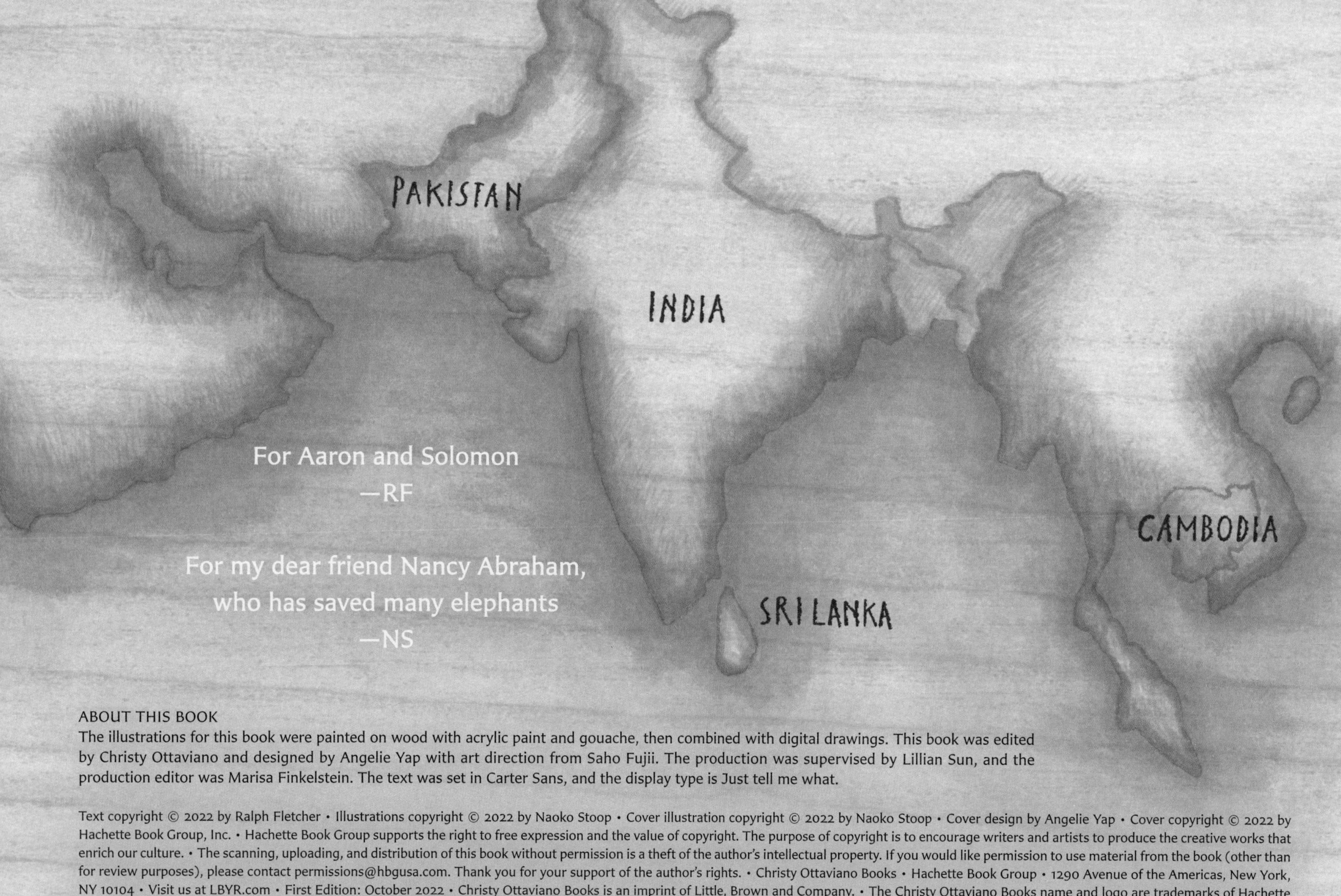

For Aaron and Solomon
—RF

For my dear friend Nancy Abraham,
who has saved many elephants
—NS

ABOUT THIS BOOK
The illustrations for this book were painted on wood with acrylic paint and gouache, then combined with digital drawings. This book was edited by Christy Ottaviano and designed by Angelie Yap with art direction from Saho Fujii. The production was supervised by Lillian Sun, and the production editor was Marisa Finkelstein. The text was set in Carter Sans, and the display type is Just tell me what.

 • Christy Ottaviano Books • Hachette Book Group • 1290 Avenue of the Americas, New York, NY 10104 • Visit us at LBYR.com • First Edition: October 2022 • Christy Ottaviano Books is an imprint of Little, Brown and Company. • The Christy Ottaviano Books name and logo are trademarks of Hachette Book Group, Inc. • The publisher is not responsible for websites (or their content) that are not owned by the publisher. • Photographs on page 32 copyright © Aamir Qureshi / Getty Images. Photograph on page 33 copyright © Tang Chhin Sothy / Getty Images. The title page is page 1. • Library of Congress Cataloging-in-Publication Data • Names: Fletcher, Ralph J., author. | Stoop, Naoko, illustrator. • Title: The world's loneliest elephant : based on the true story of Kaavan and his rescue / Ralph Fletcher ; illustrated by Naoko Stoop. • Description: First edition. | New York : Little, Brown and Company, 2022. | Includes bibliographical references. | Audience: Ages 4–8 | Summary: "A heartwarming picture book featuring the true story of Kaavan the elephant, his unlikely bond with musician Cher, and his rescue by veterinarian and animal rights activist Dr. Amir Khalil" —Provided by publisher. • Identifiers: LCCN 2021051510 | ISBN 9780316364591 (hardcover) • Subjects: LCSH: Elephants—Anecdotes—Juvenile literature. | Captive elephants—Anecdotes—Juvenile literature. | Animal rights—Juvenile literature. • Classification: LCC QL795.E4 F54 2022 | DDC 599.67—dc23/eng/20211104 • LC record available at https://lccn.loc.gov/2021051510 • ISBN 978-0-316-36459-1 • PRINTED IN CHINA • APS • 10 9 8 7 6 5 4 3 2 1

The World's Loneliest Elephant

by **Ralph Fletcher**
Illustrated by **Naoko Stoop**

BASED ON THE
TRUE STORY OF KAAVAN
AND HIS RESCUE

Christy Ottaviano Books
Little, Brown and Company
New York Boston

This is the story of Kaavan, an elephant who made a remarkable journey from one part of the world to another.

When Kaavan was a baby elephant, he lived at an animal orphanage in Sri Lanka.

In Pakistan at the time, the president's daughter dreamed of owning a pet elephant. He arranged to have Kaavan brought to his country as a birthday gift for her.

His daughter was thrilled. "Can I keep him here at our house?"

"We don't know how to take care of an elephant," her father replied. "Kaavan should live at the zoo. You can visit him there."

They sent the little elephant to the Marghazar Zoo.
Kaavan was one year old. He would live at the zoo for the next thirty-five years.

In the summer months the zoo got very hot. Kaavan was kept in a very small enclosure with no trees or bushes. He was chained most of the time. His handlers made him perform for visitors and beg for money.

One time, Kaavan got an infection.

Another time, he came down with a fever.

But he rarely got the medical treatment he needed.

Luckily, Kaavan had a mate, a female named Saheli.

The two elephants lived together at the zoo for twenty-two years. They had a special bond.

But one day, Saheli became ill.
Her condition grew worse and worse.

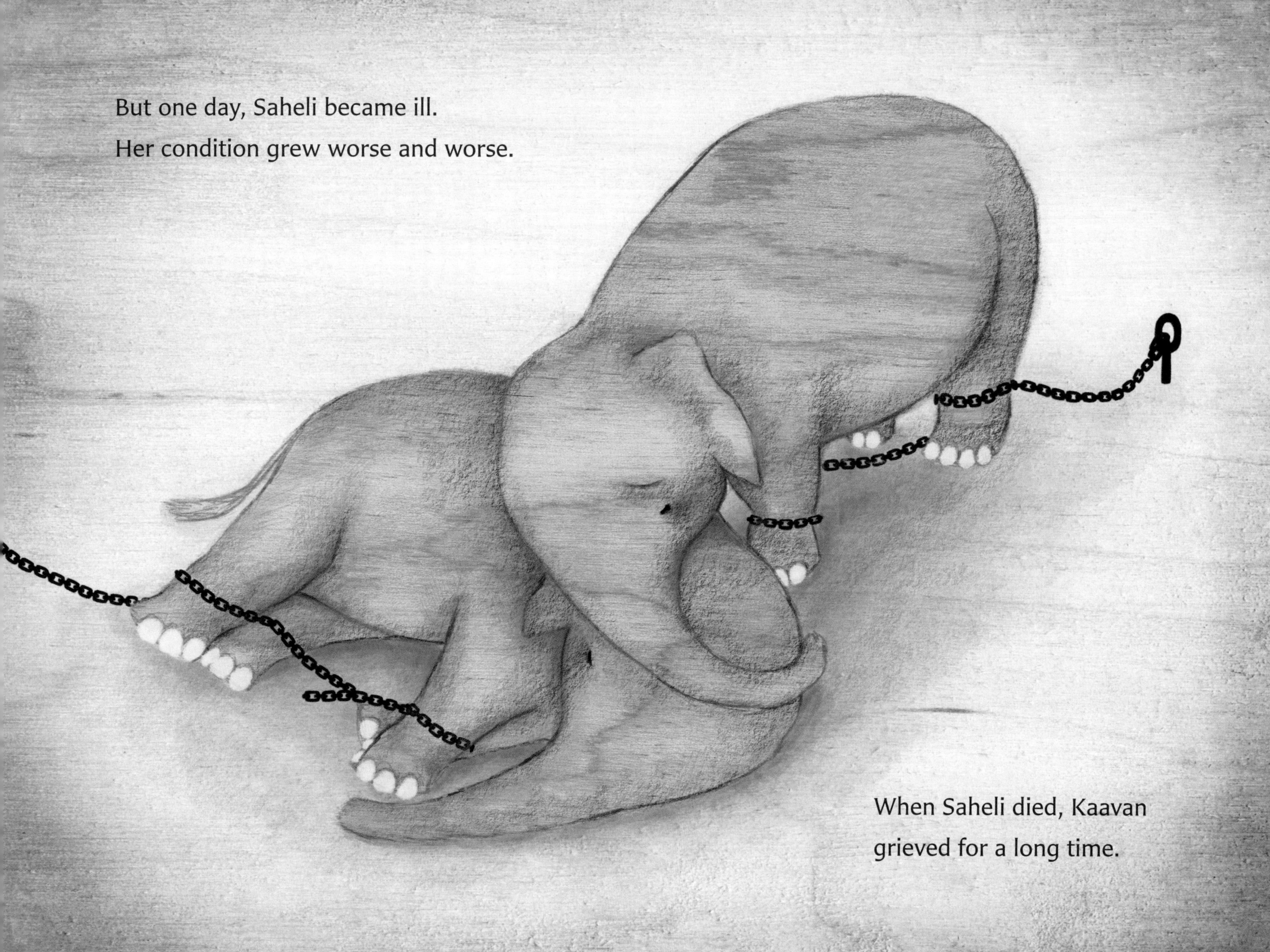

When Saheli died, Kaavan grieved for a long time.

Now Kaavan was alone. People called him the world's loneliest elephant.

An Egyptian veterinarian named Dr. Amir Khalil heard about Kaavan and went to visit him. The elephant seemed depressed and in poor health. He behaved aggressively toward Dr. Khalil when the veterinarian tried to approach him.

Dr. Khalil went to the government of Pakistan and told them he was concerned about Kaavan. He asked the government to release the elephant from the zoo and move him to a better place.

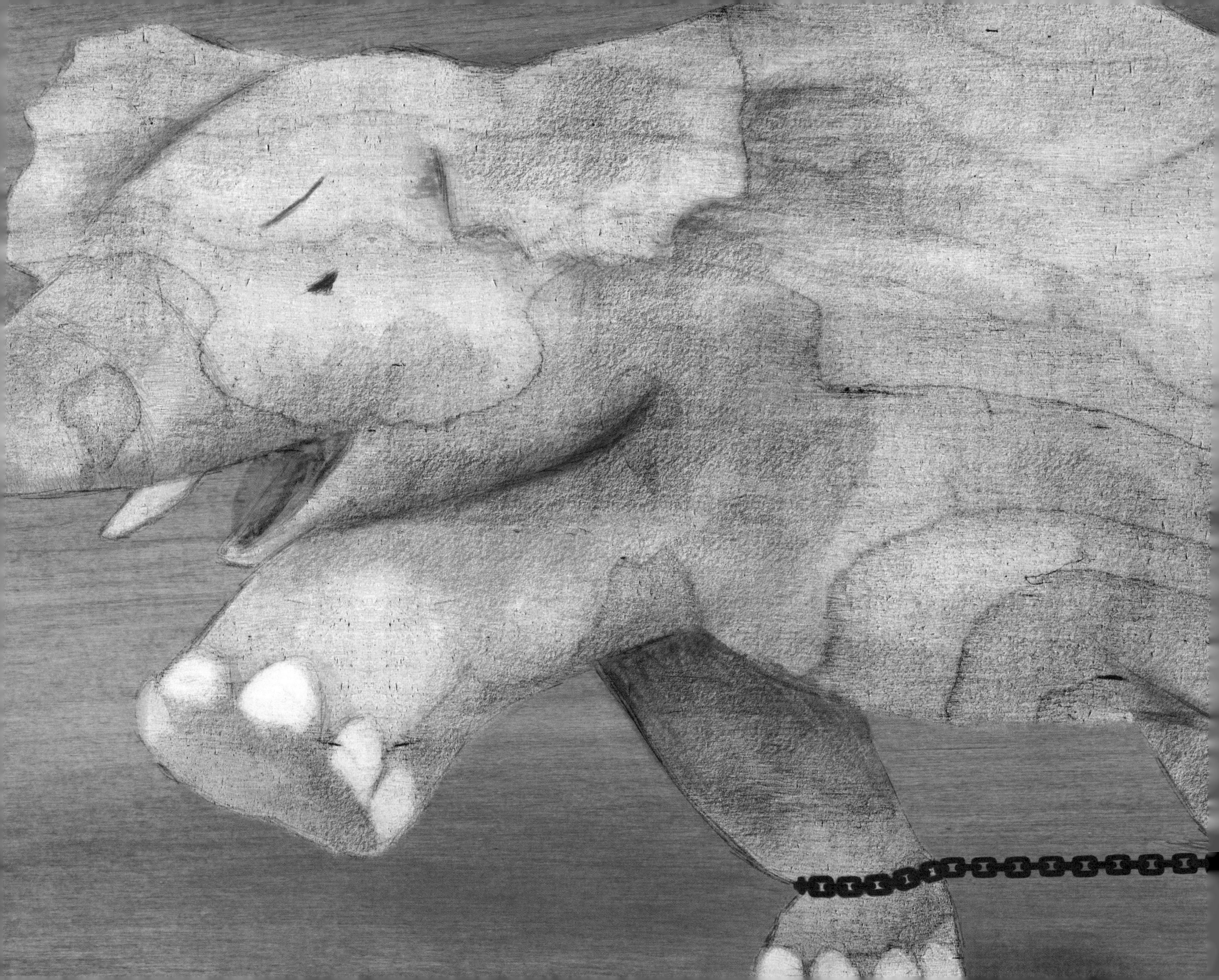

Other groups took up Kaavan's cause. Animal rights activists posted videos on social media. One group started a petition demanding that Kaavan be released. More than four hundred thousand people signed the petition.

Cher, an American entertainer and activist, learned about Kaavan. She and her organization, Free The Wild, joined the efforts to have Kaavan released from the zoo.

In 2020, the Islamabad High Court ruled that the Marghazar Zoo should be closed and all the animals in the zoo relocated. The High Court put Dr. Khalil in charge of finding a new home for Kaavan.

Dr. Khalil had learned about the Kulen Promtep Wildlife Sanctuary, a spacious wildlife community in Cambodia. It seemed like a perfect spot for Kaavan to begin a new life. But how could an aggressive bull elephant be moved to another country four thousand miles away?

First, Dr. Khalil needed to earn Kaavan's trust, and that wouldn't be easy. The elephant was suspicious of people.

One day, for no particular reason, Dr. Khalil started singing to Kaavan. The elephant appeared to enjoy it.
"He seems to like songs sung by Frank Sinatra," Dr. Khalil said. "Especially 'My Way.'"
After that day, whenever Dr. Khalil visited Kaavan, he made sure to sing some Frank Sinatra songs.

At the zoo, Kaavan didn't have an opportunity to exercise, so he had grown overweight. Dr. Khalil put him on a healthy diet. Kaavan needed to slim down for the big journey ahead.

The diet worked. In three months, Kaavan lost about one thousand pounds. Now Kaavan weighed only nine thousand pounds.

Moving an elephant from Pakistan to Cambodia would cost a great deal of money. People from all over the world helped raise the funds.

A special crate had to be constructed to transport Kaavan. Dr. Khalil began giving Kaavan his meals inside the crate so that Kaavan would feel comfortable in it.

On November 24, 2020, Pakistan's current president Dr. Arif Alvi and his wife visited the Marghazar Zoo to say farewell to Kaavan.

Five days later, Dr. Khalil led the elephant into his travel crate. It was lifted onto a large truck. Kaavan left the zoo for his new home.

Workers at the airport helped load Kaavan and his crate onto a huge airplane. An hour later, the jet and its famous passenger rose into the sky.

Dr. Khalil accompanied his friend on the flight. He could tell that Kaavan was feeling anxious, so he sang some Frank Sinatra songs. That calmed Kaavan down.

The flight from Pakistan to Cambodia took eight hours. When Kaavan's plane touched down in Cambodia, the airport was packed with people.

Hundreds of journalists, photographers, animal rights activists, and friends had gathered to greet Kaavan. Monks came to bless him. It was a joyous celebration.

On November 30, 2020, Kaavan took his first steps into the Kulen Promtep Wildlife Sanctuary.

Now he had plenty of room to roam and play.
And he had three female elephants
as neighbors.

Dr. Khalil was there to watch Kaavan and make sure he would be okay.

"This morning was the first time his trunk touched another trunk," Dr. Khalil said. "It was very touching."

Whenever Dr. Khalil visits Kaavan, he never forgets to sing.

And when Cher comes to visit, she sings to Kaavan, too.

Kaavan is enjoying his new home at the wildlife sanctuary. His favorite pastimes include taking long mud baths in his own personal spa,

playing with new toys,

and eating meals with his elephant friends.

The world's loneliest elephant is lonely no more. And that's a wonderful thing.

Kaavan: Yesterday and Today

In 1985, when Kaavan was a baby elephant, he was taken from his home in Sri Lanka and shipped to Pakistan. He was going to be a birthday gift from the president of Pakistan to his daughter.

For nearly thirty-five years Kaavan lived at the Marghazar Zoo in Islamabad, Pakistan. After the death of his female companion, Saheli, Kaavan lived in isolation. Most of the time he was kept in chains.

Many people became aware of Kaavan's situation, including Dr. Amir Khalil, a veterinarian from Egypt, and Cher, the entertainer and activist from the United States. Dr. Samar Khan, who at the time was a veterinary medicine student, visited the elephant and started a petition demanding his release. She was joined by a large community of Pakistanis who grew deeply concerned.

Kaavan's plight went viral on social media and news sites; he became known as "the world's loneliest elephant." People from all over the world began working to get him released and bring him to a better place to live.

On November 30, 2020, Kaavan left the zoo in Pakistan for a new home. A huge cargo jet transported the elephant to Cambodia. Today, Kaavan lives at the Kulen Promtep Wildlife Sanctuary, in Siem Reap.

Kaavan at the Marghazar Zoo in Islamabad, Pakistan.

Dr. Amir Khalil soothing Kaavan while he is in his travel crate.

The Kulen Promtep Wildlife Sanctuary has set up a jungle area of thirty acres where Kaavan can roam. Although the sanctuary continues to provide fresh fruits, vegetables, and banana tree stems at the end of the day (and sweet potato for breakfast), Kaavan gets most of his food from the jungle: grasses, bush leaves, and fruits. Those who see Kaavan on a daily basis report that he is enjoying his surroundings. The Kulen Promtep Wildlife Sanctuary even built a spacious new pool for Kaavan.

Kaavan has been alone for many years and is not used to sharing a space with others. The animal experts at the Kulen Promtep Wildlife Sanctuary use a "gradual release" model, which allows Kaavan to slowly adjust to his new home.

Kaavan enjoying his new home at the Kulen Promtep Wildlife Sanctuary in Cambodia.

In July 2021, two female elephants, Arun Reah and Sarai Mia, were introduced to Kaavan inside his enclosure. The three elephants stood shyly side by side. This was the first time in many years that Kaavan had full contact with other elephants. Kaavan and his companions are taking their time getting to know one another.

Kaavan endured many difficult years, but his admirers can take comfort in knowing that he now lives in a more sustainable home, where he can enjoy a much healthier and happier life.

SOURCES

Ashley Westerman, "'He Will Be A Happier Elephant': Vet Describes What It Was Like to Rescue Kaavan," *NPR*, December 5, 2020, https://www.npr.org/2020/12/05/942214125/he-will-be-a-happier-elephant-vet-describes-what-it-was-like-to-rescue-kaavan.

Cambodia Wildlife Sanctuary, accessed August 27, 2021, https://www.facebook.com/CambodiaWS.

Cher, *Cher and the World's Loneliest Elephant*. Directed by Jonathan Finnigan. Smithsonian Channel, May 19, 2021. Documentary, 46:00. https://www.smithsonianchannel.com/details/show/cher-the-loneliest-elephant.

Free The Wild (website), accessed August 27, 2021, https://www.freethewild.org/about.

Jackie Salo, "'World's Loneliest Elephant' Kaavan Finally Makes A Friend," *New York Post*, December 1, 2020, https://www.nypost.com/2020/12/01/worlds-loneliest-elephant-kaavan-finally-makes-a-friend/.

M Ilyas Khan, "Kaavan, the World's Loneliest Elephant, Is Finally Going Free," *BBC News* (Islamabad), November 28, 2020, https://www.bbc.com/news/world-asia-55060433.

"Rescue Kaavan," Four Paws International, updated August 2021, https://www.four-paws.org/our-stories/rescues-success-stories/rescue-kaavan.

Shamber Alexander (videographer), "Saving Kaavan, the World's Loneliest Elephant." Produced by Aisha Mir. SCMP Films, *South China Morning Post*, June 27, 2021. Video, 14:01. https://www.scmp.com/video/scmp-films/3138696/saving-kaavan-worlds-loneliest-elephant.